God Barking i[illegible]

And Further Glimpses
Of Animal Welfare

by

Eilleen Gardner Galer

Text & Photography: **Eilleen Gardner Galer**

Editor: **Gregory J. Kroitzsh**

Photographic Editor: **H. Donald Kroitzsh**

Assistant Editors: **Geeta Anand**
Virginia Popko

Published by:

Five Corners Publications, Ltd.
HCR 70 Box 2
Plymouth, Vermont 05056
USA

God Barking in Church
And Further Glimpses of Animal Welfare
ISBN: 0-9627262-7-3

I think I could turn and live with animals, they are so placid and self-contain'd,

I stand and look at them long and long.

Walt Whitman

1819–1892

Talking about animal welfare. . .

Lisa

. . . and listening.

Tip

Sir!

Chapter I

The humbly eager young man was quickly hired that busy morning after only the most cursory interview, for we were desperate. With alacrity he set about fitting his meager belongings into the living quarters provided at the shelter. Nothing told us that he was dyslexic.

Came evening and our new night attendant sallied forth on his first response — to a place of worship disturbed by a dog. On the call slip he noted:

"God barking in church."

It was choir practice night. And although the church fathers had urged participation by even those capable only of "making a joyful noise," Fido's contribution was unbearably off-key. So a quick call was put through to us and our lad kindly removed the offending pup.

But that was not the end of it. Little did the choir director dream that right then, under his very nose, a torrid love affair had flamed. One of his singers, unable to forget the personable little waif, fretted all week lest an owner be found, and hadn't a moment's peace until she happily adopted him. Afterward, the dear lady became so engrossed in her pet that nothing else mattered.

A fellow church member unburdened himself to us: "Now she's got a dog, she won't sing in the choir. Somebody's got to talk to her. She's our best soprano."

Although our plot of suburbia borders on the infinitesimal, things happen all the time. Birds and squirrels, even raccoons, get into chimneys. And a lady reported a bat in her bedroom. Our night man was not sure that he "had jurisdiction over bats in ladies' bedrooms."

Rescue of animals in distress is a major concern of the humane society. Scared cats up tall trees and dogs fallen into abandoned wells or trapped in the storm sewer must be safely brought down or up. No less urgent was our care of the suffering duck kept chained on a blistering apartment house roof, and the red-tailed hawk tied to a fence post at an auto repair shop. Nor can we forget the gentle, trusting goat kidnapped by three boys for target practice with their new bows and arrows.

Less routine was the plight of a pair of lion cubs left three weeks without a keeper after their

Such elegant table manners.

owner, a rodeo performer, had been gored and killed by his Brahma bull. Chained in a flimsy trailer parked on a nearby farm, the animals were neglected until neighbors became alarmed and notified the authorities. Their call for help brought us into the case.

We found the cubs slithering about in filth, with only leavings of spoiled food and bone-dry drinking pans. Our first thought was water. From the farmhouse well we carried full buckets up a long hill. While the cubs drank, we managed a tricky bit of mucking out with the only tool handy, a loose-handled rake. It was generally agreed that if the rake came off, we would not go into the cage after it. A substantial meal restored good feeling. At ease, the male enjoyed a little neck scratching as he pressed hard against the bars.

Very early Sunday morning, through streets all but deserted, the trailer with its occupants was towed in to the Washington, D. C. Zoo for safekeeping. And later through purchase the pair became permanent residents.

In animal welfare there is no foretelling what may come to pass. It could be rounding up sheep escaped from a Christmas nativity scene, or counseling an apartment house manager aghast at finding "a dead reindeer tied up to the play-ground swings."

No matter how improbable the summons, it becomes the Society's urgent responsibility to calm the discomposure caused by a "dragon" in the box hedge, or a "wildcat" under the hood of a car. The dragon proved to be an iguana, the wildcat an oil-smeared, terrified red Persian.

And towards midnight of a world sheeted up in snow and ice, an agitated woman shouted into the telephone, "There's a big white bear sitting on my porch. Come and get him!"

Calmly waiting there under the porch light was our old friend the Great Pyrenees. An inveterate roamer, he regularly got lost, poor fellow. From numerous odd-hour pickups we were so familiar that our driver had only to point toward the truck and the huge dog obediently clambered up onto the seat beside him.

Since Arlington has long been a big-city suburb, any farms are ancient history. But our night man was awakened in the wee small hours:

"You know what?" a man's voice queried. "There's a MULE in my yard."

Shocked, our lad whispered, "Are you sure?"

Talk, talk . . .

"Yep. I just looked out, and there he is!"

Clutching at straws — "You ain't been drinkin' ner nothin'?"

"Nope. Sober as a judge."

"Well, what do you expect me to do about it?"

"Come and git 'im!"

Give the lad credit. He got one of his friends out of bed to drive our truck while he rode the animal bare-back to the shelter.

My role in animal welfare has included sharing the pride and sorrows of many devoted pet owners:

"You don't know who Ernie is, but I'll tell you. He's a real charac-ter. He doesn't realize he's a dog. He thinks he's people. And my husband rottens him to death."

Or the bulging wallet album is unfolded to pictures of the snowy Angora, and a withered face smiles wanly as the aged woman recounts the story of her cat's significant life. "She never had a flea on her little body. She was angelically clean. Her name was Celeste and we talked to her like she was somebody. She loved to sit in my sister's lap. My sister was tall and had a lap. I'm too short and fat. I don't have a lap." The voice trailed off. "They're both gone now, and I'm all alone . . ."

Contents of the portfolio are not always predictable. The lady who adopted our clubfooted kitten spoke of her neighbor boy, "You know cat fanciers and their kitten pictures. Dog owners are as bad. Bird people are no trouble at all. But Bobby with his newt is something else."

While it may be true that bird fanciers seem not so inclined to carry photographs, they are proud to tell stories: of the parrot staunch of faith who announces to the world, "Jesus loves me", and the parakeet they must sing to sleep each night with "Rock-a-bye-Baby." No lame brain was the parakeet so smart that he remembered not only his name but also his address and telephone number. Such cleverness brought him safely home when he flew out an open window and was lost for a whole day.

Less conventional pet owners in moments of sorrow need sympathetic understanding. Parting with one's "nice friendly" alligator or losing a pregnant horned toad can be just as devastating as loss of the usual good friend.

A war veteran who had been helped through convalescence by a spider told me: "I had a pet black spider named Pete. I used to watch him just to pass the time. One day I saw a small red spider spin a thread down the door frame. He came down so far, then he went back up, leaving the web hanging there. While the little red spider was busy up in the corner, along came Pete and he climbed down the web. Immediately the little red spider came down after Pete, who was three times his size. I thought, 'Here's where Pete makes a kill.' But do you know, before you could bat an eye, that little red spider killed Pete. He just curled up like he was frizzled. I blame myself. I sure missed the little feller."

Sad too was the solemn moment when we accepted a wee turtle named Jimmy, ailing beyond recovery. And a small boy handed up to the counter a little wooden cross that he had made for the grave of his departed friend.

Tranquility

Thistle

Chapter II

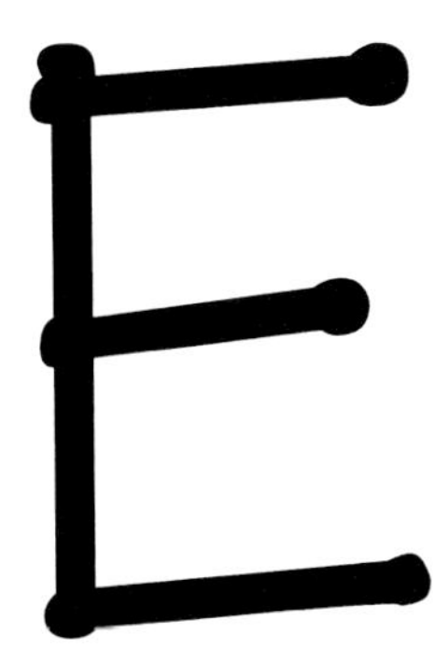very day for nine years she had packed her husband's lunch and the dog's lunch and watched them go off to work together. Since puppyhood their little Cairn Terrier had taken his responsibilities seriously, as behooves an important member of the family. When not accompanying the breadwinner, he considered it his duty to assist her in entertaining. Politely he stood ready to shake hands with each arriving guest. And when everyone had been graciously received, he would bounce up into his personal armchair where his keen darting glances missed none of the fun.

"Joey thinks he's people," she said fondly.

Zeke, our mynah bird, knew for sure. When rebuked for being an old blabbermouth, he told everybody, "I'm a people with feathers."

We who love animals agree. And for the enlightenment of those who may be of vehement contrary opinion and tempted to ridicule our "sentimentality," I would point out that dictionaries supply one meaning of the word people as animals collectively: "living creatures, especially of a certain kind, as the bee people, the ant people." It follows then, on the highest authority, that there are cat people and dog people and bird people. And as the saying goes: "it takes all kinds of people to make a world."

In my life's haphazard drama, those who have made the world a better place are the furred and feathered folk, and the people who cherish them. With the former, during twenty years of volunteer service in the animal shelter, association ranged from speaking acquaintance to enduring friendships among the approximately two hundred thousand creatures who have come and gone. While cats are my first love, along the way I met some very special dogs, and little birds have pleasured me endlessly. With the human kind, our common devotion to animals has made possible the felicities of an engrossing career.

I got into the humane society quite innocently. An invitation to their annual meeting, followed by a proposal that I prepare their press release for the occasion, hardly seemed the beginning of a life sentence. Prior to the meeting they had eased my concerns about two stray cats that had

Saturday Night Bath

appeared on my doorstep. And I gave them a small donation.

So, they found out about me. And from taking my money, they took my life.

My friends and associates in this sometimes sweet, sometimes bitter bondage are people of superior sort, naturally sensitive to animal woe anywhere. They think nothing of handing over their lunch to some hungry stray wandering the streets, or leaving a cozy bed, however late the hour or foul the weather, to go in search of the helpless, lost or injured. And if need be, in response to citizens' concern about an elephant shackled on hot concrete, visit the zoo and feel an elephant's hide in order to judge the condition of the hapless pachyderm.

Little things, as well as big, are taken in stride, like baby-sitting tadpoles, or foster mothering bird orphans, even to restaurant dining with a basket nesting a clamorous *youngster* that must be regularly fed (a scandal to the waiter).

Besides the rescue of animals in distress, the return of lost pets to their owners is one of the more pleasurable services of humane work. No effort is spared to bring about a happy reunion, always expedited by any kind of identification.

Fortunately the footsore little Cockerpoo still wore a legible link with his past. My friend Dorothy, champion saver of weary wanderers, spotted him for a lost dog running with another in the street towards midnight as she returned from the theater. But the flow of traffic made it impossible for her to pick him up.

Next morning, however, she found the same animal shying around her office parking lot. And though it was January and snowing, she patiently sat outside with a bite of food and friendly persuasion until she managed to collar the skittish waif. Once caught, he gave in completely and slept from sheer exhaustion the whole day under her drafting board.

Because he had a bad cold, she took him for treatment, which the doctor felt came just in time. The small vagabond's luck had nearly run out.

From his collar dangled the rabies tag of a California veterinarian. When called long distance, the good man could hardly believe his ears, that some kind person was phoning not from around the corner but from Washington, D. C. something like three thousand miles away. The dog had been missing since October.

I just came from the hairdresser

Immediately came a call from his overjoyed owner asking that her pet be put aboard a plane for home. And there his rapturous reception at the airport made news headlines.

How the little fellow got safely across the broad United States remains a mystery. His family surmised that his ardor for hippies led him to run off with one of the groups departing from a nearby beach. Somebody must have watched over him on the long journey, but his escapade left the traipsing pup a frightful distance from home and loved ones. Only the compassion of a stranger saved him from an early death in the gutter. That jaunt evidently cured his wanderlust. And subsequently Dorothy had the pleasure of visiting the little rascal at home in the bosom of his family.

Another waif who wore no clue to her family connections could not be returned. Several mornings in a row commuters observed a small black dog lingering by the roadside in a wooded gully near the river. When Mildred heard of this, she got in her car and went searching. But two trips were necessary before she caught a glimpse of the shadowy creature. And it took considerable coaxing to develop reciprocal trust.

Never a trace of the dog's owners was found; yet she gave evidence of gentle upbringing. Once acquainted she beamed her friendliness in all directions, responded obediently, and would sit up to shake hands when introduced. For this she was chosen from some forty animals in the shelter to be the first pet of a nine year old girl whose friends gave her a puppy shower. Gifts included such accessories to creature comfort as personal comb and brush, several fancy collars, a variety of doggie toys and tasty tidbits.

The mislaid pet best identified and most decorated was a pert little black and tan Fox Terrier found tied to a heap of baggage left unattended at the airport. He wore a tan leather collar and a new harness to which was affixed his owner's name with Marine Corps insignia, an American flag, one buddy poppy, and a cardboard sign: SPECIAL SALE IKE Sale Price $250,000.^{00} My Best Friend.

Potted

Chapter III

The number of homeless animals whose destinies I have directed would be impossible to guess. But they and their families must be legion, memorably peopling the dim yesterdays. Concerning some, I recall only snatches of conversation.

When his wife and three daughters chose a female dog, I said to the husband, "With so many girls in the family, you're kinda outnumbered."

Serenely he replied, "We have a bird at home and he's a boy."

A tall, lean husband with a tiny, plump spouse cheerfully answered my question as to whether there were children in the family, "No, just adults, children at heart. And idiots about cats."

When I cautioned a cheerful black woman to take good care of her new little puppy, she answered, "I'll treat him just like my own blood baby."

I remember the young couple who came in to adopt a dog because, the wife said, "My husband has an aquarium, and I can't get cozy with a tank of fish."

Among others, the picture recurs complete. An adoption possibly unique occurred one dismal day when a car rolled up to the shelter and we could see framed in the near window a frowsy small black head with a band of scarlet below. I remarked, "Somebody's coming in a red sweater."

But sweater girl remained in the car. Her mistress entered alone. She came to find a kitten. Her toy poodle Petite had been grieving since losing her pal, a friendly big neuter of the neighborhood who had disappeared. Petite cried till she developed sore throat. And the vet said, "Well, why don't you get her another pal?"

Most eligible at the moment was our adorable little longhair. But the lady questioned whether she might not be too old at eight months to adjust to a dog.

I suggested that we bring them together. Petite arrived in her owner's arms. I set our dainty miss on the counter top. Straightway she crossed over to rub her head against Petite's chin. She buzzed, the poodle quivered with excitement, and Mama said, "Looks like that's our girl."

I have my eyes on you

Then, consulting the kitten's record, I found the notation, "loves dogs." Not every day is choice of a new cat entrusted to the family dog. Not every day is a dog-loving cat waiting.

Saturday placement hours were often so crowded that shelter visitors must take a number and be served in turn. On one particularly frantic weekend a stranger waited nearly an hour in the busy office, seeming in no hurry, and allowing everyone to leave ahead of him. He was a homely man with small eyes and flat nose, and neither his crew cut nor the dun-color turtleneck sweater improved his looks. Beside him on the bench a kitten had been left temporarily in a carrier, and he gently stroked its striped brow with the tip of his finger.

When at last the place emptied, he came forward smiling and introduced himself with the hope that he might adopt a pet.

That he was a settled bachelor, a kind and understanding man, I already knew from friends who had recommended him as very worthy of our confidence. And today the kennels offered a wide choice between Chihuahua and Great Dane.

"I am not looking for a beautiful animal," he said. "I am not beautiful myself." His glance took in the Russian Wolfhound watching us through the window. "Perhaps you have an orphan nobody else wants."

We had indeed— a dingy mop of a dog named Daisy who for two weeks had been desperate to be noticed by somebody. *Anybody.* But she was so lacking in canine graces that visitors ignored her pleading, Due to privation and puppies her shape somewhat resembled a lumpy sack. Her coat was thin in places and pied, white with mud-colored patches, and her eyes didn't match, one being blue, its mate brown. We had remarked of poor eager Daisy that she would have to make up in personality for what she lacked in looks.

She *had* to be this man's desire. And after a tour he requested an interview with her, though his mind must have been made up, for one could hardly imagine a homelier pooch. Brought into the office, she made straight for the gentleman. He lifted her beside him on the bench and put the question: Would she like to go home with him? And Daisy was thrilled to the tip of her ratty little tail.

During the business of adoption papers she pressed close to him, rolling her eyes. He kept an arm securely around her. And when he crossed the room to give us his signature, she trotted right

One tooth left

at his heels. Until safely out the door with him, little Daisy was plainly fearful of being left behind.

Within the month he brought her back. But only to show her off. He was proud of her glossy coat and pearly teeth, both of which he brushed each day. But mainly he dwelt upon her fine intelligence and ever-charming ways. Plainly they adored each other. "It was a real love match," he said.

I had a few love affairs of my own through the years. Some I look back on with satisfaction, since I was able by my personal diligence to make life happier for these creatures on the mercy of the world, which too often is sadly lacking in mercy. Others, because my best was yet not quite good enough, haunt me still.

Constant temptation in the presence of so many appealing animals can be borne only through the utmost restraint. Emotional involvement in humane work soon leads to an insidious undermining of all resistance as your sympathies draw you in deeper and deeper. So many need you, and it is so easy to lose count, that soon you have a home full of the homeless, with no room for the worthy ones yet to come.

You may find yourself in the predicament of a kind lady whom I met when I visited her place to photograph the injured goat that her husband had rescued.

Having pleasurably greeted and been welcomed by many of her animals, I said, "There's nothing like pets."

She replied, “They’re people.”

And as generally happens, the more you live with, the more adjustments you may be forced to make. She spoke of the problems with their dogs. They had eight, all waifs for one reason or another. And sometimes, when they barked or howled with the fire sirens or they fought among themselves, and all must be fed according to individual preference, some on one side of the house, a trio on the other, and a couple indoors, there was talk that they couldn’t keep so many.

So, another home must be found for . . . Not Baron — he’s sheriff and patrols the place. And Hans would never be happy anywhere else. Bibi sleeps with the youngest daughter. Bernard’s skin ailment requires treatment. And Bruno has a thing about rocks. They got him in the first place because he took the neighbor’s rock garden stone by stone and deposited it in his own yard. So the count reaches eight. And nobody leaves.

Vera, describing another humanitarian, said, “She was a wonderful person. She moved to Ohio. It was like the ark. She had old goats that were crippled, old horses had cataracts, eleven cats, nondescript, old and nobody wants. She took them all. I never met the woman, but I like her.”

For the unwary volunteer in animal welfare there is ever-present danger that the heart may overrule the head. One must early and sternly draw the line, keeping in mind that it is utterly impossible to help every one. For when your own dwelling place is bulging with waifs, there will still be more. Many, many more, sadly.

And so we live in hopes of matching our precious pet with the perfect family. The purring palaver of some adorable cat, or the boundless affection of a lovey-dovey dog, spurs us to wide searching for just the right home place.

In a patch of pleasant memory lingers a little gem. A large and charming foreign family, having all been kissed in turn by an enthusiastic small black dog, agreed unanimously that they couldn’t live without him. And great was their excitement when they prepared to take him home. Above the bustle of herding her joyful brood out the door, Mama announced, “I get him a red collar.”

Grown-up daughter demurred. “I thought you buy him jewelled collar.”

“Later jewels,” Mama said firmly. “A red one to begin. Red goes good with black, don’t you think?” And observing the dog’s happy departure, she added half to herself, “I hope he likes music . . . when Papa plays the flute.”

Mr. Frog comtemplating his domain.

Chapter IV

In foreign travel I have found some unusual pictures, and rare, such as scenes among the brooding ruins of Anghor Wat in the jungles of Cambodia, now lost to warfare. Yet none of these has brought me such rewards in their power for good as the animal studies I have caught right here at home.

During my twenty years of volunteer service with the Arlington Animal Welfare League, our shelter accommodated something like two hundred thousand animals. And among that varied throng I found a wealth of pictures with charming stories to match. Always I enjoyed complete freedom in choosing my models, as well as unlimited time for photographing them. During one year I brought home with me eighteen different parakeets. And I long ago lost count of the kittens I borrowed.

Furthermore, I could expect to be notified whenever anything picture worthy occurred. After a Kentucky Warbler flew into a window pane and knocked himself out cold, I was alerted to be present when he recovered his *joie de vivre*. Singing ecstatically, his long pale claws clutching an evergreen branch, he made a picture rated outstanding in exhibition. Besides illustrations for the League's annual reports and occasional newspaper publicity, my photographs furnished irrefutable evidence needed for prosecuting gruesome cruelty cases.

But my chief contribution to animal welfare lay in developing our humane education program. This was a photo-essay, of the sort that must be conceived and designed with imagination, intended to promote kindness and understanding. It was widely shown in schools, scout meetings, civic associations, camera clubs and wherever else we might be welcome.

From the shelter's multitude of guests, I chose a variety of actors. It is important to impress kids with the people traits of animals: if you treat your pet as a playmate rather than as a plaything, he may sit up and shake hands, the kitten may arrange a daylily bouquet, and the cat may laugh at your jokes. One little lad asked, "How does a cat know that a joke is funny to laugh at?"

The appeal of a box tortoise savoring delicious pink watermelon moved a child to whisper, "He's purty."

It is always interesting to hear about the children's personal pets. We begin by asking how

Flower Arranging

many have dogs? Cats? Others? There is a bunny or two and a duck. One little black boy raised his hand. His pet was one guppy.

Personable goats, a mother and daughter named Eloise and Rosemary, made their stage debut (thanks to me) in a high school production of Heidi. This event, duly publicized, went off without a hitch, except that the teacher's students gave her a beautiful bouquet, and Eloise ate half of it.

Among the wild creatures we introduced to our delightfully receptive audiences were the tame gray fox with a passion for soft drinks, and a sociable little skunk named Penny who routed an unsuspecting bill collector, then obligingly waddled from hand to little hand so that an entire kindergarten might pet her. And not forgetting the exotics, we included Samantha, the kinkajou, who liked to wear things on her head. As a gift for her first birthday party, I bought for her a shocking-pink doll's bonnet with satin ribbons that tied under her chin. The fetching little comic was a smash hit!

Over the years, our program was enthusiastically received by more than twenty thousand persons of all ages, from tiny tots to the elderly in retirement homes. A power for enduring good were the converts, among them: thirty pre-schoolers whose mothers reported that their pets had "taken on new status," and the camera club gentleman who told me, "I don't care for animals, but I like yours." Inspiring rapport with our furred and feathered friends may prevent much thoughtless cruelty.

The program was great fun with a serious purpose. Mindful of the old adage, "as the twig is

bent" etc., I have sought to influence the impressionable young through a memorable experience and thereby nip in the bud any evil tendencies that might later create horrors. Such, for example, as the teenager who beat an old Poodle to death, and in court for cruelty to animals admitted that he had abused and killed two kittens, two ferrets, a bird and his Boxer puppy.

Samantha the Kinkajou

To develop concern that builds active participation and solid financial support, the public must be made aware of the tremendous scope of animal welfare work. Small humane societies are usually volunteer organizations of animal lovers performing, without thought of remuneration, a vital community service. For needed funds they must depend upon the generosity of an oftentimes fickle public. Burdensome demands and insufficient income are generally chronic. A humane society accepts responsibility for sheltering unwanted animals, rescue of those in distress, investigation and prosecution of cruelty cases, return of lost pets to their owners, and careful placement of shelter guests in approved homes.

These services are full of drama, with a tremendous cast of characters. The play may be poignant or comic. It is never dull.

People must be educated to a high regard for the wonderful creatures who share our world. Revelations through pictures may open eyes that only half see and minds that half the time don't think.

Jebby

Chapter V

At the time I began my apprenticeship in animal welfare work, Mittens, our first beloved cat, was a matron of eleven summers. And already she had presented us with numerous progeny. Such a performance I would not tolerate today, conscious as I am of the multitude of animals born just to die. Our subsequent felines have all been spayed or neutered early in life, which makes for happiness all around. Raising endless litters of kittens for casual distribution, I sadly learned, is one sure way of contributing to the miseries of this world.

But in Mittens' day the protective shots were not yet perfected, so that surgery presented a fearful risk. We lost four of her kittens after spaying from what was then called "cat typhus." This experience led us to delay with our old girl until she passed her tenth birthday, when we had to chance it on account of her troubles.

Mittens was a little magpie, which is to say black and white, though not, as an acquaintance put it, "one of those cats patched and pied who look like they've been thrown together and all their ancestry is showing." She was jet black except for snowy vest and small feet and long white whiskers.

The week after we moved into our new, still-unfinished house, Mittens became a member of the family. A most important one, for she introduced us to the joys of cat company, which we have not since been without.

I had not the remotest notion of adopting a pet the day we visited a friend's farm and saw a black mama cat with her frolicking eight-weeks-old kittens.

Incautiously I remarked, "They say it's good luck to own a black cat."

Our friend seized her opportunity. "You want one?" she said eagerly. "I can't keep them all."

Looking back I am amazed at my ready acceptance. Coolly I selected. "I'd like that one with the white feet."

Mine was in some ways a happy ignorance in those days. Had I given thought to the sex I preferred, I very likely would have acted according to the dictates of others and chosen a male,

Caught in the act!

because "males are less trouble." Actually they are not. So I would have missed the joy of an engaging feminine personality. By the same token we would have avoided the problems of Mittens' family life. Although she never had large litters, usually singles or twins, the little ones came regularly, and all sleek and fat, for she was a devoted mother. We enjoyed the kittens; yet their futures caused us great concern. Less than a dozen found good homes through my personal efforts; the others were an imposition on the humane society.

Work in the animal shelter quickly taught me the terrible wrong of such permissiveness. Humane societies constantly wage a dismaying battle against the folly of indiscriminate breeding. Many more young are born than there are good homes waiting. Today, with only charming spays, I need not worry about the fate of kittens I am responsible for.

I have no patience with the woman who snapped that she would not "mutilate a cat for our benefit." Such misguided persons are guilty of shocking cruelty in permitting the birth of countless animals for whom they cannot guarantee a good life. They get rid of helpless creatures to just anybody, and never worry a tittle what happens to them. The possibilities are heart breaking to think about.

One smug housewife boasted that she had "gotten rid" of 250 kittens! Hearing her, an animal lover horrified at the abuse of a kitten used as a football kicked about by several children, considered such wrong-headed individuals beyond all redemption. "Whatever their other good and virtuous acts," she said, almost in tears, "they can never atone for this evil."

Though Mittens' next to last kitten was stillborn, it gave her no trouble. Six months later she produced a lone magpie who grew to robust manhood, our boy Jebby.

We called him Jeb Stuart because of some quaint conceit that he and the Civil War soldier had mustaches alike, although in fact Jebby was the negative of the image, his being white on black. But people found his name amusing, so that visitors to the cat shows remembered to ask for our boy year after year.

Jebby was his mother's last born to survive, a magpie like herself with ermine paws and chest, but also white patches on either side of his nose. We never had a better behaved youngster. From the time he began to crawl about until he got big enough to venture out of doors, his playpen was a large carton cut low for him to see out on all sides. There he had his housekeeping facilities and favorite toys, and there his mother must have given him strict instructions to stay, for even when he

grew sturdy enough to climb over, he never left his box.

Hey . . . What is she doing?

At his station in the mainstream of family affairs the little tyke would sit with tiny forepaws resting on the edge of his enclosure, bright eyes watching all the passing to and fro, mostly of Brobdingnagian feet and ankles. He had no playmates, but his mother often engaged him in lengthy murmurous conversations with much rustling of papers that sounded like the pair might be going over their reading matter together. And Mittens always looked on with quiet satisfaction when we wiggled a string or tossed a ball for her lone kit.

It was through Jebby that I received my initiation into the local cat shows. About the time he was three months old, I found myself one of another group concerned with animals — the Cat Fanciers. I discovered them through their newspaper ad requesting the loan of cat photographs for publicity announcing their first show. Attending that show I learned there is such a thing as the *Household Pet Entry*.

At the very casual invitation of their secretary, scribbled across my envelope in which cat photographs had been mailed to them, I dropped in on the club's monthly meeting. I went expect-

ing merely to observe and perhaps meet a few cat people. The dispatch with which they inducted new members surprised me. Without so much as a by-your-leave, vote was taken on admitting me to membership, and there being no dissenting voice (how could there be, when I was a total stranger to everyone present), the president said, "Welcome, new member!" And the treasurer extended her palm, "Have you got two dollars?"

Despite my unblemished acceptance, it soon became clear from the conversation eddying around me that I was in beyond my depth. These were all breeders of fancy felines, and all well-schooled in the facts of life among high class cat circles. The talk of genetics, discussions with some disagreement about inbreeding and line breeding, all the shared knowledge and enlightening chatter left me in utter darkness.

Startling to the novice was the conversation of two women shrill above the hubbub:

"Oh, isn't that a shame!"

"What's the trouble?"

"She just had a miscarriage and lost four perfect little Siamese!"

"How simply ghastly!"

Happily it developed that I was not the only household pet owner gathered into the fold, so that my inferiority was shared during the Cat Fanciers' second annual show by three other very nice people. Between us we entered four neuters and one spay.

With Jebby, on his first birthday, in his first show, the public was intrigued. But to the judge, even though he looked his handsomest, our boy rated only second best, being nosed out by a fine black fellow named Trouble, who afterwards sent Christmas cards to his friend Mr. Jeb Stuart.

Jebby developed into a big-boned cat who settled the scales variously between eighteen and twenty-two pounds. Cat show judges, however, accustomed to weighing the merits of finely proportioned purebreds, failed to appreciate the large and plump household pets. Audiences tittered when a judge lifted Jebby and sank under his weight, but our lad's avoirdupois invariably cost him the blue ribbon. Second place, however, remained his personal rank for the duration of his show career.

To some breeders of fancy aristocrats, lowly household pets sullied the pure excellence of any

Bimbashi — Cameo Persian

show. And they gave vent to their irritation in sarcastic remarks as unkind as they were unnecessary. Trouble's owner commented plaintively that some of the breeders were "mighty catty."

The paying public, on the other hand, very much enjoyed the household pets. More than one visitor sheepishly confessed to a preference for the *ordinary* cat, yet wondered about our huge specimens, whether these were not some special breed. Proudly we informed everybody that these were only extraordinarily fine *ordinary* cats.

Jebby, then weighing all of eighteen pounds, still looked less imposing than twenty pound Butter Ball, a cream longhair. But generous were the compliments on his magnificent size, his unusual markings, his clear amber eyes.

His name brought smiling comment. One dapper individual even inquired whether I was kin to Jeb Stuart; I resembled a portrait of him in Richmond.

The sincere interest and admiration expressed by the visitors was heartwarming. But the disdain and articulate annoyance among those shepherding purebreds introduced a sour note into the proceedings for the lowly household pet owners. Next to Jebby was benched an exquisite chinchilla Persian whose owner had no scruples about informing me that her jewel had been purchased because her husband said, "It costs no more to feed a good cat than a mutt."

And when the judging ran on quite late, exhibitors awaiting the final choice of Bests felt grievously put upon having to sit tediously through consideration of the household pets.

One breeder in particular spoke her mind at every opportunity. A short, gross woman, with an ungirdled rump broad as that of a draft horse but considerably less firm, she wore at the same time felt bedroom slippers and over her mannish haircut a helmet-like hat with curled feathers erect dead

center of front, and she spouted insults through the gap of her missing front teeth.

When, following the purebreds, I carried Jebby up to the judging stand, she yapped, "From the sublime to the ridiculous!"

I stopped beside her. "Now just what do you mean by that?"

She quailed but blundered on, "I mean this is far different from your black one."

"This is my black one."

"I mean your longhair."

"I have no longhair," I replied with chill dignity.

Within seconds, another breeder, a weasel-faced woman, added her own affront. Watching Jebby in the judging, I realized that I was standing in front of her, and I apologized.

"It's all right," she snapped. "There's nothing to see right now."

Next on the stand appeared a handsome cinnamon brown neuter, a very unusual *ordinary* cat. The judge looked him over, smiling, and said kindly, "What have we here?"

"That's what we're wondering," piped up the toothless one, with a smirk all around. "Looks like a ruined Siamese."

Jebby's was a hairbreadth loss of first place. The judge debated between him and a fine gray tabby neuter incongruously called Kitty Puss. She studied them carefully, explaining, "Condition counts for a lot, and these two are in excellent condition."

Finally she awarded the Blue ribbon to Kitty Puss and placed the Red on Jebby. So, we could only try again next year.

And twelve months later we braved the barbs once more. My Charlie dropped us off at the hotel, anticipating that we would find a crowd of pet owners standing in line with fancy satchels full of yowling fury. Jebby, always an amiable fellow, accepted the shows calmly. Never was he difficult, except that he made such a heavy burden while standing in line for vet clearance. I had to squat on my heels, resting him on my knees. And strangers noticing him smiled down on us.

To our dismay, Kitty Puss's mama and I found our humble entries benched right beside the

The whole story in black & white.

toothless one's faultless aristocrats. We anticipated rough going.

But our erstwhile tormentor was a changed woman. Cordiality personified, she made an inordinate fuss over Jebby. Not a whisper of disparagement escaped her. And when a visitor inquired if he was "just plain alley cat," Madam rebuked her, "That's a Domestic Shorthair!"

Hearing others praise him, she must add her share. "Oh, he's just the cutest rascal!" she cried. "And his mama worries him to death."

She kept us laughing. Getting her precious darlings settled, she waddled about, lisping comment and advice through her vacant teeth. Still in felt slippers and feathered helmet, she busied herself all over the place. Finally her beauties were installed in a double cage lined with azure silk hangings and soft cushions, and protected from germs of the careless human by sheets of cellophane with large hand-lettered signs — DO NOT TOUCH.

But she took time to fasten Jebby's water cup with one of her pinch clothespins. And she gave me the benefit of her long experience with various remedies for all kinds of illness. At the first suspicion of sinusitis, I should dose with certain pellets, one every morning for three days. And for something else, which I neglected to jot down, liver pills were the answer. Moreover, she insisted

upon pressing into my hand a packet of tiny yellow tablets for immediate use should my "sweet thing" show signs of cystitis.

Only when she happened to be engaged elsewhere could I enjoy the visitors. One young couple had noted my name. They wanted a kitten marked exactly like Jebby. But of course so far as he was concerned that would be impossible.

A charming matron, stylish in plum-color wool with jet ornaments, paused to chat. She told of her cat that everybody spoiled, her maid most of all. He even ate corn-on-the-cob on her Chippendale sofa. She laughed stylishly "My children wouldn't be allowed to do that."

And there was the Very Important Personage in rose cinnamon knit with suede beret to match, her wrinkles brightly painted, her jeweled hands nervously twitching, who described in detail her perfect cat that the President (of Cat Fanciers) just begged her to show. "And, my dear, he just swept everything before him; he won ribbons and money; and, my dear, he was just the pride of the show. I was in bed under a doctor's care but, my dear, I wrapped up in a blanket and came down to the show and —"

Our toothless neighbor interrupted. Her Mignonette's number had been called. She patted and fluffed the languid Persian and waddled off with her perfect cat.

By show time Sunday morning, having fattened her purse with the sale of two high-priced kittens, Madam Toothless was as chummy as a schoolgirl and as silly. She caught hold of Jebby and kissed the top of his head, exclaiming how cute he was, insisting that she must order for him some good meat.

Her magnanimity soared to climax in response to praise from out-of-town visitors who had been urged by their neighbors to see Jeb Stuart especially, "Isn't he darling?" she gushed. "Look at that 'stache. I'll have to give a special trophy for him next year."

Her raptures were more exhausting than her insults.

Charlie suggested, straight faced, that perhaps I should have made her a present of the cat. Such heresy! And our boy that very moment encumbering his lap, handsome in his neat clerical garb. He put the question to Jebby, "Can you show me in black and white any good reason why I should keep you?"

For an answer he got a doting look from our big, beautiful hunk of manly feline.

Innocence *Gigi*

Chapter VI

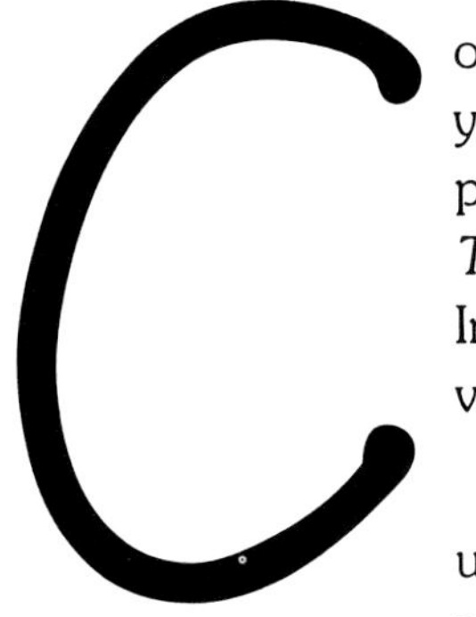

Cold rain discouraged visitors to the Animal Shelter that Saturday afternoon. Our young kennel man, who had been bitten by a terrified cat, amused himself preparing an insurance report: Machine, Tool or Thing Causing Injury — *Tomcat!* Kind of Power: Hand, Foot, Electrical, Steam, etc.— *Jaws!* Has Injured Died? — *Nope.* I happened to be on duty only because the scheduled volunteer had unexpected house guests. Benevolent Fate arranged that.

A car door slammed and a woman dashed through the downpour bringing us a kitten, just five months old, shaded silver with green eyes, the little one was exquisite. Yet this drab had twice tried to palm her off on willing kids whose parents promptly sent them back with the cat.

"I couldn't get rid of it no other way," she whined, "so I had to bring it to the Pound."

That expression "get rid of" is guaranteed to rile us. And this woman's heartlessness was infuriating. She wanted only to sign the release and leave. But we managed to piece together a sordid tale of indifferent pet ownership. It seems the family had been given a silver Persian which they did not particularly want, but merely tolerated. Since cats were never allowed in their house, the poor beauty was left outdoors year-round. And she must have been in dreadful shape.

Later on, when her daughter Gigi came to us with near fatal ear mites, the unfortunate mother so haunted me that I tried to rescue her. I promised a very good home. Her perverse owners only gave me the brush-off. They cared nothing for her; yet they obstinately refused to part with her. I never knew what happened to the poor creature. I could only do my best for her four kittens brought to us at different times.

This youngster was adorable. Hearty of purr, she danced on little tippy toes, pounced on the tip of a pencil, and gave my hand friendly little nudges with her nose. She was so lovely and loving, and finally so utterly irresistible, that in spite of my already full house, she came home with me.

On all sides she met hostility. Bluette, in the rocking chair, looked down upon her with a deep growl, menacing as distant thunder. Even so, her greeting was comparatively polite; the others squalled their insults. And the baby Lisa was so meek and demure that their rejection pierced her to

Bluette's Kitten Picture

the heart.

But growing older she asserted herself until she was the *enfant terrible* of our household. A small white spot breaking the dark stripes above her nose reminded me of the verse about the little girl with the curl in the middle of her forehead. Lisa was like that little girl. When she was good, she was very, very good, and when she was bad, she could indeed be horrid.

Innocent and inoffensive kittens who came for only a brief stay were practically annihilated by her eloquent antipathy. Her snarl was enough to split a tonsil. And when she SPAT!!, she hurled almost visible venom at those unfortunates who encountered her.

Strange felines were not tolerated even behind closed doors for so much as a brief photo session. Her savage presence could dampen the gayest party. There was temporarily boarding in our spare bathroom a white longhair mother with her one black and three white kittens. Newspapers were spread down to protect them from the cold tiles, and I could hear much rustling as they romped together. Glancing in, I discovered that they had been making merry with the toilet paper. Spread wall to wall lay yards and yards of green tissue. The little ones were having a marvelous time. But they froze and every hair stood on end when Lisa spat at them through the crack under the door.

Nor was she more tolerant of her sister Gigi. Both were beautiful, one a shorthair, the other long, and I visualized a handsome portrait of two glamour girls. But Lisa wrinkled up her nose and hissed the elegant Gigi out of countenance.

Even when our other girls, in the mood for friendliness, chirped a pleasant greeting, Lisa would snarl back. More sensible than most of us, she didn't associate with others just because they meant well.

Sometimes in her sleep she was intimidated, surely by some hideous ogre. Suddenly the fur ridged along her back, weirdly. It was enough to make my own hair prickle. I wondered what she could be dreaming.

But the demons of her slumber were never so terrifying as thunder or firecrackers. Storms could rage all around our older girls and they remained perfectly calm, but Lisa slunk away to find some deep, dark hiding place. I always tried to shield her from atmospheric tumult and explosive merrymaking.

This phobia of hers gave us a scare one bitterly cold New Year's morning. I awoke in the gray dawn to discover that she had not come up to bed. This was so unusual that I thought she must be shut in one of the downstairs rooms. So I got up and went searching. But the rooms were empty. Nor was she perched on her favorite window ledge in the basement. I opened closet doors. No Lisa. I repeated the circuit, growing more alarmed, calling her name, peeking into possible hide-aways. Still no sign of her.

We had last seen her about midnight. With a stab of fear that she might have slipped out of the house, I looked on the porch. It was empty. But Charlie could have opened the back door to check on the weather. I snapped on the back light. There in the snow *little footprints went around the house!*

I had awful visions of my pal wandering five long hours in the bitter cold. Perhaps she was even frozen stiff. The temperature hovered around twenty degrees, so that the rhododendron leaves hung tightly rolled and pencil-thin. I grabbed a wrap to go searching.

At that moment she nonchalantly appeared in the cellar door. From the mustiness of her fur, she must have been far back in a corner behind boxes and sleeping so soundly that she never heard my calling. At breakfast we solved the mystery. Last midnight's New Year's firecrackers had fright-ened her; so she hid away in the deepest part of the house.

Now, unperturbed, she breakfasted well, then took an interest in the young starlings with speckled feathers fluffed, sitting silently on our wall waiting for a handout. Expectant also, the

Fury *Lisa*

mockingbird perched on our neighbor's chimney, scanning the landscape, while above him the heat waves he sought in winter distorted the dark, bare oak branches.

In the bloom of their youth both Lisa and Gigi did their bit for animal welfare through appearances in the cat shows. We in humane work see so much of the neglect and cruelty suffered by cats that we seek to elevate them in the public esteem. Adoptable beauties from the shelter appeared among the household pet entries to impress upon people the fine qualities of *just ordinary* cats available and waiting for a loving home.

Lisa's show career was neither as long nor as brilliant as Gigi's. In her first appearance, however, she triumphed. The judge smiled on her, commenting aside to the assistant, "Good head . . . cobby . . . good body." The other agreed, "And how!" Lisa topped off her collection of ribbons with fancy rosettes and trophies in Shaded Silver Shorthair and All Breed: *First & Best of Color,* and in Specialty: *First, Best of Color, Best Household Pet Spay & Best Household Pet.*

Although she was a belle, Lisa's second appearance became her last. All because some nasty male sprayed the side of her cage. With that, the fat was in the fire. She had a fit. Absolutely beside herself, she snarled and snorted, refusing to be calmed until we got safely home. Show biz was not

for her. So we gave up public glory for private tranquility.

Those of the opinion that cats are not affectionate should have known my Lisa. She was loving as any dog, always with matchless grace and dignity, and constant beyond the capacity of many people. For twenty-one years she gave me her single-minded devotion.

Lisa was a unique little personage. Besides the loveliness of her shaded silver fur and green eyes, and the charm of her character, she had keen intelligence. Between us there was perfect understanding. And our long companionship brought rare happiness. Other people she could ignore. Cats she loathed! I was her whole world. Utter contentment meant just being with me. When we were both young, she considered it her duty to waken me each morning. She began with inquiring little chirps, then purring softly, she peered into my face. When I pretended to be still sleeping, she would reach out a velvet paw to pat my cheek.

She liked to share my second cup of breakfast coffee, taking a sip that was mostly cream from her own little cup plate. Whatever the day's urgencies, we always lingered awhile, sitting together. I told her how beautiful she was, stroking her striped brow or giving her a little kiss on the bridge of her nose. She responded with ecstatic purring and, when asked, would give me a little kiss on my cheek.

When she was gone, she left an almost unbearable void. Wherever I gazed into emptiness, I saw her. She was gone, yet she was everywhere, so completely had she entered into my life during her brief span.

We shared simple pleasures. It was delightful sitting on her rug spread over the grass, above us the sunny oak leaves and patches of blue sky, about us the songs of birds and the fragrance of homely single roses hardly larger than a twenty-five cent piece. Folded close beside me, she purred softly, her bright glances following every movement in the shrubbery. If I leaned toward her, she gave me a little kiss on the forehead, or cheek, or the tip of my nose. When I took her out a treat and she left a dab of liver paste on the plate, we watched the daddy longlegs lap it up like delicious chocolate ice cream.

She communicated in charming little ways. As a rule she was eagerly responsive to my wishes. But when she must refuse something, she did so apologetically. Yeast tablets, for instance, were sometimes acceptable, and again, not. She would sniff and draw back, then rub her whiskers against the pill, a cute gesture that plainly said, "I know it's good for me, but not right now."

Lisa with a few of her show ribbons.

Time, that only endeared her to me the more, sped by. Admired, loved, Lisa had a serenely happy life. But as her birthdays approached the equivalent of one hundred for a human, the years began to take their toll.

At the age of seventeen she had to have all her teeth out. But their loss never reduced her prodigious appetite. She simply gummed her food or swallowed it whole. Unlike humans, her face never looked caved in.

She developed a sinus infection that gave her the sniffles, requiring medication and somebody to wipe her nose. Deafness closed her in. Unless she was nearby, I had to shout and clap my hands. Yet deafness had its compensations. She no longer suffered through the tumult of thunderstorms.

Cataracts clouded her vision. But her personal night lights helped her to find her way on nocturnal journeys.

Age wizened her. And the stairs got steeper week by week. When something above was on her mind, she plodded up without fuss. But sometimes the terrible height seemed to daunt her. Then she sat at the bottom, mournfully wailing. And I carried her up.

The old girl accepted her infirmities with remarkable fortitude. But on one occasion she rested her forepaws against my shoulder, her clouded eyes searching my face, and murmured, "Mmmm?" Meaning, "Can't you wish away this cruel spell that binds me?"

It was only in her last few weeks, well into her twenty-second year, that she failed rapidly. From ten pounds of cuddlesome curves in her prime, she wasted away to little more than skin and bone. Her fur lost its sheen; spit and polish took more spit and energy than she could summon. Her purr, once so joyfully vigorous, became muted, almost inaudible. She talked to herself when alone.

Yet her appetite for good food and the life around never diminished. Though frail of body, she remained strong of spirit. She still craved our sweet talk over morning coffee. When I praised her beauty, probably my voice no longer reached her, deaf as she was. But she understood. Love penetrated her failing faculties. In token of her abiding affection she would give my cheek little nudges with her nose and gently squeeze my arm where her paw rested.

She was, if possible, dearer in her decrepit old age than in her radiant youth. She was so patient, so utterly trusting. Calmly she accepted her infirmities, depending upon me with implicit faith.

To save her needless exertion, I fixed her comfortably in a small upstairs room with her housekeeping facilities handy. And I carried her food up to her. But one evening she decided to join the rest of us for dinner in the kitchen. Though weak and wobbly, she made it down the stairs and gamely climbed back up.

Next morning her back legs were paralyzed. She seemed in no pain, only withdrawn and peaceful. Resigned, perhaps, that the end was near. *She knew.* Dreamily she gave my hand the usual loving nudge with her nose. A hearty breakfast quickly disappeared. But she could not stand. I worried that she might later be in pain at an hour when professional help was not available. A sad decision had to be made. It seemed kindest to preserve her dignity and save her from possible anguish. I made an appointment with the hospital where she was known and admired. In the vein of her frail little paw the vet gave her an injection, and she went painlessly to sleep. He said, "She had a good life."

Mai Ling

Chapter VII

Some people show remarkable talent for finding clever names to suit their pets. But tastes vary so. I've been trying for years to persuade somebody that Paleface should be the name of a white cat. I've yet to find any enthusiasm for my idea. Nor was my suggestion favorably received when Jan consulted me about the naming of her expensive little number, a flame-red longhair kitten that cost seventy-five dollars. She had consulted the dictionary. How did I like Ahtina? "It means Red Indian," she said.

But one should not endorse the first offering. Besides, the subtlety is lost if the meaning must be explained. She asked me to give it thought. I did. Later I told her, "You want an Indian name. How about Minnihaha?"

"Minnihaha?"

"Minnie for the cat and haha for the seventy-five dollars." Very clever! But of course there again the subtlety had to be explained.

Naming our pets has entailed considerable research and vehement argument. Some people manage to be so original: brother and sister kittens born in a pantry became Potsy and Pansy. Twins, impossible to distinguish, responded to Hither and Yon. A Maltese who stretched down and down when picked up was Long John Silver. I like Cunningham for a cute clown— cunning ham. Nor can I forget two boys as tiny kittens who were called Large and Small, but when they grew up, Small was larger than Large!

Commendable for her foresight was the young women who adopted a cat despite strict rules forbidding pets in her apartment. She named the puss Goldfish. Then if the management questioned whether she had any pets, she could truthfully answer, "Just Goldfish." She was evicted anyway.

A teenager brought to the shelter a stray black cat she had found. She said his name was Izzy.

"If he's a stray, how do you know his name?" I asked.

She giggled, "You know— Izzy yours?"

Stuffed Shirt

Of Dorothy's two cats, her Col. Mosby arrived in the arms of a child who, making room for a new puppy, appeared at her door saying, "You can have my cat."

Mosby, a rich brown tabby and white, with green eyes, developed into a sleek and handsome neuter who won fancy gold-fringed rosettes big as plates and numerous silver trophies during his fling in the cat shows.

I never heard how Col. Mosby's name was chosen, but it caused some raised eyebrows. It seems that Dorothy's skin bruised very easily. During a routine check-up the doctor discovered black and blue spots on her behind.

"What are these bruises?" he asked.

Replied Dorothy, "Oh, that's where Col. Mosby bit me!"

It's just that the poor fellow got *so* hungry. While she cut up his meat he nibbled her ankles. Then he would stand up, nibbling higher and higher. The less prompt the service, the more damage he managed to do.

I had gone for an afternoon call upon a friend living in one of the large apartment house complexes notoriously inhospitable to pets, and was surprised to find there on the steps of her building, surrounded by a band of screeching wild Indians, a cat remarkably calm. He was young, a neat brown tabby, wearing white on chest and throat snug as a fresh turtleneck sweater. He sat erect, black tail curled over his white feet, facing the frenzied children pell mell into a bloodcurdling war dance. He followed their antics with lively interest. His *sang-froid* was amazing. But they made *me* nervous. And I deemed it wise to rescue one so unaware of his peril.

At my greeting he rose in polite response, quite agreeable to joining other congenial company. Together we entered and presented ourselves at my friend's door.

She laughed. "Well, who is this?"

"A kindred spirit."

With perfect poise he accepted our invitation to enter, and was pleased to receive light refreshment in the shape of a dish of milk, which was the best that the house had to offer. Though young, he behaved like a man of the world. There was no ill-bred snooping about the apartment. Instead, he joined the gossips' circle, relaxing on the rug at our feet, napping when our chitchat palled.

Soothing Syrup

After he had caught up a little on his sleep, he arose and stretched, with a yawn you'd think might unhinge his jaw, and asked to go out. Since the bedlam had subsided, we humored him.

An hour or so later, when I opened the door to leave, there he lay, asleep on the welcome mat. Somebody had been persuaded by his courteous request to hold open for him the heavy entrance door, and he was not at a loss to find the portal to hospitality.

The sudden appearance of this charming fellow was mystifying. Where did he belong? How did he happen to be there?

My friend was entranced. She wanted to keep him. Forthwith, I was pressed into service as cat sitter *just over the weekend* until she could make inquiries and, if no owner appeared, get permission to adopt him.

Nature Study

Ever a pushover in such delicate dilemmas, I let myself be taken in by friend and furry stranger alike. Nobody claimed him, and the management was adamant: pets in the apartments were not merely frowned upon, they were strictly forbidden.

Such unreasonable restrictions troubled our self-assured feline not at all. He accepted the situation gracefully, making himself comfortably at home with me for just over a decade.

When I opened the car door that day, he stepped in with all the nonchalance of one accustomed to a chauffeured limousine. I shudder to think of the consequences had some cruel person taken advantage of his delightful insouciance. His manner was that of a very intelligent equal. He was a very gentlemanly cat.

Arrived home he preceded me up the steps and so was first to meet the boss whom he

If you treat your pet as a playmate,
he may laugh at your jokes:

This was a funny story.

This was an even funnier story.

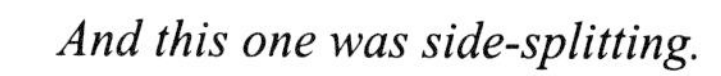

And this one was side-splitting.

greeted with a cheerful, "Blurrp?"

Charlie said, "Who the hell are you?"

Quickly I explained the very temporary nature of his visit. We called him Zeppelin because he was lighter than air. But once he got his belly full and flesh on his bones, he never weighed less than sixteen pounds. Before he got too solid and comfortable, his reflexes were sufficiently quick that he could swoop down from the woodpile on a weasel, which I am told is quite a feat, the latter being lightning fast.

But Zeppy was not a hunter; bird lovers would appreciate him. He never to my knowledge caught a bird. Even when the mockingbird dived at him peacefully minding his own business, he showed great forbearance, only nattering petulantly. He was rather a student of the wildlife round about, observing with interest the eating habits of Possum Pete. Jason was another interesting oddity. Zeppy kept his eyes on this curious specimen, his head jerking with each hop. If the toad rested, he prodded it with his paw. This bit of nature study ended disastrously when he got too pushy and gave Jason a little nip, for which he got a bitter dose that left him spluttering. After such treatment he lost all interest in toadology.

As a flower fancier Zeppy had no equal. Like our other cats, he must always sniff any bouquets I cut. But he also strolled through the flower beds, lightly touching with his nose this posy or that. Handed a Daffodil, he reacted sensuously, lying on his back and hugging the blossom to him, so vigorously inhaling its fragrance that the fragile plant was left limp and broken.

Zeppy was an accomplished yawner. In later life, yawning was his favorite form of exercise. His yawns were wide, wider, widest, for a look deep down his little red lane. This propensity made him one of our star performers in the humane education program. To dramatize the message that pets should be treated as play*mates* rather than play*things*, there is a homely pooch shaking hands, and a kitten arranging the flowers. He may even laugh at your jokes! Zeppy laughs at a funny story, an even funnier story, and one that's side splitting!

After a group of thirty pre-schoolers gave rapt attention to our animal friends, responding to Zeppy with screaming delight, their mothers reported that their pets had "taken on new status."

Though hundreds of children have responded to Zeppy with wild abandon, only one serious little lad questioned the cat's mirth. "How," he asked, "does a cat know a story is funny to laugh at?"

Lordy! The troubles I've seen.

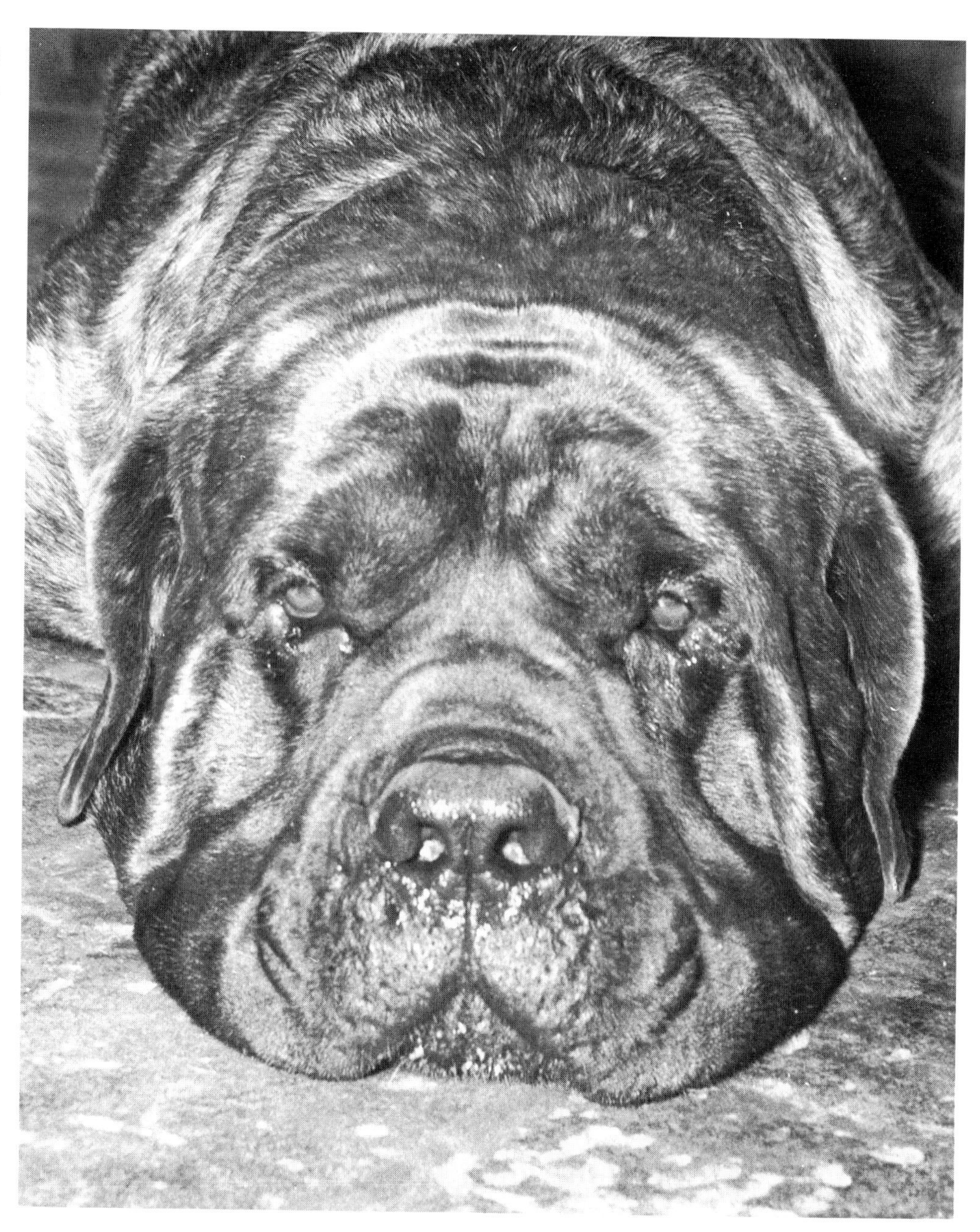

Chapter VIII

A situation requiring courage and delicate diplomacy has been my adoption by a dog, and I in no position to reciprocate. One must be made of stern stuff to hand over to strangers the homeless pet who makes it quite clear that he or she prefers to belong to you.

Yet knowing that the animal will have a better life with others, I have been compelled to ignore this infatuation of the moment. I steel myself, therefore, against the partiality for me shown by the long-neglected little Yorkie whose bright eyes follow my every move. With us, he would be one of many, surrounded by feline animosity, sharing divided affections. The little fellow deserved a home exclusively his own, there to nap undisturbed on his owner's bed. On mine, too many cats taking their ease leave only the skimpiest room for me.

And though bandy-legged, freckle-nosed Bascomb the Basset was a photographer's dream, the soulful way he rolled his eyes and trod on his silky brown ears; yet I could not provide ten country acres and lively youngsters for romps to keep him well and happy.

So, too, I could have lived nicely with the Basenji, since the barking of dogs has never been exactly music to my ears. An African breed of ancient lineage reaching back to Egypt in the time of the Pharaohs, Arthur possessed admirable qualities, being fastidiously neat and highly intelligent. Though barkless, he communicated.

I admired the proud little dog, and well he knew it. We got on famously. A short snappy walk was a treat to break the monotony of his day in the kennels. To be invited as helper a while in the office put sparkle in his eyes. But this indulgence led to complications, since Arthur was more interested in me than in the visitors who sought to adopt him.

Twice, with my blessing, he went home with nice families. As many times, he came back because he was so unhappy. Arthur had definite ideas about his proper place. So when next I appeared for my stint at the shelter, there sat my little friend waiting to greet me.

At last came a lady who fell for Arthur in a big way, dreamed of him overnight, and returned to make him hers. I was not present for that event. But this home evidently satisfied Arthur's

Chow Time at the Shelter

desires; so he did his part. And when his owner brought him back for a social call, his bearing, I'm told, was very proud.

Where some wonderful dogs come from is a mystery. They suddenly appear, alone, lost, and kind strangers rescue them from their plight. That they can be such handsome animals, so sweet-tempered and well-behaved, yet nobody ever comes to claim them is very sad. Showing all the polish of polite upbringing, these are bound to be someone's pets. Where are the kind people who have molded their characters?

Three days before Christmas a small spaniel wandered up to a man and his young son at their car in a supermarket parking lot. With night coming on, and the weather raw, they could not just leave her there on the cold concrete. So they brought her to us by way of their apartment, where the whole family fell in love with her.

Yet two months went by, and the little dog in the shelter still waited for someone to claim her. I can't imagine how I happened to miss her all those weeks, since I always made the rounds looking for possible adoptees. But she somehow escaped my notice until our night attendant left a note: "Please place #344. Come on now. Lobby!"

Seeking out this special dog, I discovered a graceful toy spaniel, white with black patches. She had a small pointed face and the *earmarks* of a Papillon, which is to say, large ears suggesting the wings of a butterfly, papillon being the French for butterfly. Bright dark eyes studied me, her long fringed tail just slightly fanning her back. Such elegant little dogs appear in old tapestries and the portraits of noble French ladies which were considered incomplete without their *dwarf spaniels*.

Marie Antoinette is said to have been an ardent admirer of the Papillon; while Madame de Pompadour had two. Highly desirable for their temperament and character, the Papillon is also extremely affectionate.

Our little waif being so obviously French, it seemed appropriate that she be called Mimi, which pleased her. Brought into the office, she captivated everyone. While I groomed her with comb and brush, and visitors admired, she stood proud as a small girl in her best party dress. And if you said, "Let's dance," she would waltz prettily. She was remarkably obedient. Show her exactly where you wished her to sit, and there she sat decorously until given permission to leave. She had been beautifully brought up.

Everyone thought her enchanting, so dainty and vivacious, and so smart. Requests multiplied for this little charmer who had gone unwanted for two long months. Applicants underwent our usual interview and scrutiny, then were sent home to await our decision. Finally the choice narrowed to an extremely nice family living in my own neighborhood. And I undertook to deliver Mimi to them.

We rode together in the car like old pals, and she trotted into my house as confidently as if this were indeed home. When she had polished off the last crumb of her supper plate, I placed her in a dining room chair where she could see me through the doorway as I dined in the kitchen. Her interest in my doings never relaxed for a moment.

And by the time we had shared a chair through a couple of television programs we two were deeply enamored of each other. Then came the cruel moment of parting.

Her new family greeted her with joy. But Mimi clung to me. She sat on my lap while we chatted a bit to ease the change over. After a while she seemed calm, even drowsy, so the mother suggested that I hand the dog over to her, then leave quickly.

But when she held out her arms, Mimi GROWLED at her!

This made the transaction all the more an ordeal. But the kind mother was understanding. She said, "Just put her down. She'll come around."

And happily she did. Mimi joined a loving family of which she was such an important member that she shared formal photographs of her mistress. New joys erased the memory of our brief encounter. Oh, fickle one! When next we met, she barked at me!

Cocker at Shelter Door *Spotty*

From the foregoing it should be apparent that only uninformed people look down their noses at animal shelters because "they have only mongrels." In reality there are few pure breeds that have not at some time had a representative of purest mold enjoying our humble hospitality. But we are not always able to recognize royalty in our midst.

What we took for a tatty little poodle was in fact a rare and extremely valuable Maltese, a toy dog for more than twenty-eight centuries an aristocrat of the canine world. It is recorded that "queens of old served out of golden vases the choicest of viands to their little Maltese." Our shelter being unequipped to provide such royal service, poor little Wags ate his portion out of the same battered bowls as our less exalted guests.

The stray we puzzled over, finally listing as a *mixed* German Shepherd, was indeed a purebred Keeshond, three times a champion and worth all kinds of money.

People often ask, "Where do you get your animals?"

Some are released by families with good reasons and bad. Strays we gathered from the streets. And one little dog came to the shelter on his own, not just once, but twice. He was a

sociable English Cocker named Spotty who, though going on four, had not developed such inner resources as would keep him happy through long hours at home alone. A family man, Spotty needed folks, and the house was just too damned empty with everybody gone all day. So the little fellow got into the habit of wandering off, and one day he was picked up by the dogcatcher. Through the license on his collar his owners were notified and they came for him.

But conditions at home were unchanged. Spotty still found the place unbearably lonesome. So a few days later, traveling through dangerous traffic, he returned to the shelter. Again his people redeemed him.

But now his solitary life was unendurable. So back he came. Once more he made the hazardous trip through clogged streets, and we discovered him peering through the shelter door, wagging his stumpy tail.

Spotty's persistence paid off. His owners released him. He was placed where somebody was home all day, and that made him so happy that he never wanted them out of his sight. After his picture and story appeared in the newspaper, he was invited to make a guest appearance on television. And his young master wore out a pocket full of clippings showing them to friends and strangers alike.

Our very strict placement policies surprised people. Those inclined to casual ownership complained that "It wouldn't be any harder to adopt a baby." But the truly kind appreciated our solicitude, commenting, "We didn't realize you were so particular."

A lady whom I guided through the adoption of her first cat, a beautiful silver tabby, invited me to tea. She said, "You have given me a wholly new conception of humane societies. I never expected such personal interest."

One query guaranteed to rile us is, "Have you got any dogs you want to get rid of?"

To which we make tart reply, "We are not interested in getting rid of our animals. We want them to have good homes."

Not just any home — only the best.

When one man was told that we would check the home, his indignation nearly choked him. "What for?" he sputtered, "it's my home. If it's good enough for me, it's good enough for the dog."

Sad Sack

Not so. Furthermore, he wanted an "outside dog." And no drafty dog house in a wintry back yard satisfied us.

Complete records on all owner-released animals served as guidance in establishing new relationships. Mocha, our friendly Mastiff, so big and ugly, but *beautiful of spirit*, came from a family with five children and went to a new home with a like number. Said their mother, "People think we're crazy — five kids and that great big dog — but she's so protective. The baby sits eating a cookie. He gives the dog a bite, then he takes a bite. She looks after him." Mocha slept in bed with some of the children, she all sprawled out, her *bed buddies* crowded to the rail.

One of Vera's placements with a Chinese family was a spectacular success. Papa called her, jubilant, "Thank you, Missy. He very fine dog. He already bite my brother!"

So great was his satisfaction that she must come, bring all her family, and have dinner at his restaurant.

Careful placement should center around the wife and mother. After all, final responsibility for the pet's welfare rests with her, and if she dislikes animals, what sort of life will the innocent lead? It has happened that a husband brought home, against his wife's wishes, a delicate little Italian Greyhound, and while he was out of town, she let the dog starve to death.

So, if mother sulks while husband and children enthusiastically discuss pet adoption in the shelter, nothing doing. The right woman for our animals exclaims, "You have four legs. Therefore I love you!"

The conversations of prospective pet owners can be enlightening. A rough looking but gentle-spoken countryman was anticipating the pleasure of owning a dog. "I figger I'll git me a bloodhound," he announced. "Then when I'm a-feelin' low I can look at that sorry bugger and feel better right away."

A bystander spoke of the Bloodhound's great intelligence.

"Oh, I ain't proud," replied the other. "If he's smarter'n me, I won't mind."

Another man boasted that his pet, adopted from us, was an animal of "super intellect." And he confessed that when his relatives came up from South Carolina, the dog was so much more intelligent that he felt ashamed to introduce them.

Bathing Beauty

Chapter IX

Among the notable characters of my years in animal welfare, one of the most articulate was a clever bird named Zeke. In his inimitable fashion he expressed the sentiments of all true friends of animals.

Zeke was a bright alumnus of the Animal Shelter, having been found footloose in a church parking lot by one of our junior members who recognized him for what he was — a mynah bird and not a local starling. She took him home and spent the next hour trying to teach him to talk back to her father. Zeke wisely played dumb.

Now a fellow so handsome, in glossy black with iridescence of purple and green, sporting yellow wattles along his cheeks and jaunty orange beak, and with a tongue in his head, naturally has the knack of winning friends and influencing people. Within the week, when no owner appeared to claim him, he was adopted by knowledgeable bird fanciers. Settled in his new home, Zeke's smart headpiece and willing practice developed his considerable eloquence and expert mimicry.

His vocal talents were in full play one afternoon during vacation time when I found him boarding with Mae. Taking the air in a shady spot, he kept up a steady chatter, his ladylike voice repeating, "Are you a talking bird?"

Finally Mae said to him, "You're an old blabbermouth."

In deep masculine accents Zeke retorted, "I'm a people with feathers."

The boy knew his stature. And he could speak for all our shelter guests. To us they were personalities. And we insisted that others think likewise. When we placed a pet in a new home, we expected it to be treated like folks.

On our preferred list are people like the couple with two young sons who adopted Champ, our noble Collie. They phoned us to say, "We've got our third boy. He has his own bunk bed."

And the young man who bought a step stool so that his little Cocker could easily reach the high top of his antique four-poster.

With my friend Lillian togetherness had long been a way of life. We were scarcely introduced

Bird Watching

when she told me about her cat, the "old man" of their household, adding complacently, "Don't tell anyone, but Yellow Boy eats at the table. He's a gentleman about it. He has his own place mat."

Real close family was what we sought.

Zeke's new home was happy for a time. Then his perfect echo of voice and mannerisms became unbearable to his owner when her husband suddenly passed away.

Back with Mae for keeps, he became self-appointed speaker of the house. He shared quarters with another mynah, small and less mouthy, a cluster of parakeets, and a Mourning Dove in temporary residence, recuperating from a back injury.

Zeke himself kept fit for all kinds of modern derring-do. Besides other interests, he actively participated in the space program, judging by his frequent announcements, "Astronaut Zeke in orbit!"

One so determined and energetic deserved a moon landing. Actually he got beyond the talking stage one evening when he escaped out the aviary door. Truly in orbit at last, he intended to stay. Mae tried to catch him; he flew above her head. She coaxed. Not even his favorite delicacies tempted him within reach. Try as she might, none of her enticements brought him down to earth.

At dusk he settled in a nearby tree. And though she worried, it seemed best to leave him alone. Soon darkness swallowed the solitary bird. She kept a light burning through the night so that

he could find the house in case he was disturbed.

Daylight found him still perched in the tree. She called out, "Zeke?"

Cheerily he answered, "Hello, Honey!" And down he came, the exhilaration of outer space just a passing fancy.

It would never have occurred to me to go out and buy a bird, with four cats in the house. But the afternoon I dropped by our shelter to photograph a little black dog who smiled, showing all her pearly teeth, I found there a woeful yellow parakeet huddled disconsolately on the perch of a beat-up old cage. He wore such an air of one miserably lost and forlorn that I offered him lodging *temporarily* until he found a ride to the country for permanent residence.

The delusion and snare of that *temporarily* idea I fully recognized where cats are concerned, Zeppy's brief weekend with us having lasted a decade. And Thistle, the weakling kitten who seemed close to her final hour, was still our girl at the age of eleven.

But with cage birds, from slight acquaintance, I lacked rapport. It never entered my mind that I might not be immune to their charms. Heretofore, all cage birds arriving at the shelter were sent directly to Mae. Now that she had moved sixty miles away, other immediate arrangements had to be made. So, I got a bird *temporarily*.

Rather, I got a bird for life. Because he gave me little kisses. And any girl would adore kisses from a fellow so handsome.

Before he had been in the house twenty-four hours we were completely captivated. Once through the door, he shed his gloom and commenced to chirp. Presently his happiness welled up in delicious little trills. We soon realized that we would miss this blithe spirit. And so Dickey took his uniquely individual place in the family circle.

We thought him incomparably handsome and one of birddom's most engaging personalities. His trim plumage was the loveliest buttercup yellow with shell markings faintly shadowed. His scissor wings and long slender tail were ivory tipped. Either side his curved beak the feathers grew like old-fashioned muttonchop whiskers with a dab of white in the middle. Bright dark eyes gave him a look remarkably knowing.

On his arrival all the cats woke up, naturally, and took an unholy interest in him. He bore their

Kissin' Cousins

Dickey & Visitor

scrutiny with amazing calm. I let them look at him from a distance briefly. After that, any undue show of enthusiasm for him was sternly rebuked.

The older girls, Lisa and Thistle, evinced from the first a disposition to be quite trustworthy. But Mai could not resist the urge to sit beside his cage and gaze upon him. Not greedily, just fascinated. While Araminta, still young and flighty, obviously was at a loss to understand why all that light meat and dark meat and gravy makings should go to waste. Her approach was the more covetous. But when, all eagerness, she pressed her face against his cage bars, Dickey gave her a peck on the nose that set her back on her haunches. That settled her.

Despite the sorest temptation, I know, all our girls quickly got the point, that Dickey was a member of the family, too. They accepted him with rare grace. And they never nattered at him the way they did towards the wild birds outside the window.

As with all our pets, we observed and enjoyed Dickey's little quirks and mannerisms. For one so small, he conveyed feeling in a big way. We kept him near us and paid him a lot of attention so that he never felt neglected or lonesome. And his joyfulness cheered us the day long.

In the mornings when he was awake I removed his cover and said, "Hello!" He yawned, and s-t-r-e-t-c-h-e-d, one foot back and the pretty wing extended, then his other foot and wing. After that he lifted both wings together up over his head like the angels in pictures.

Next he took a sip of water to wet his whistle. But before touching his seed cup he hopped down to tell his pal in the mirror all his marvelous dreams of the night before. His recital was a gentle melodious trilling, very well-bred, unless punctuated by scolding or emphatic statement. When at times his confidant seemed inattentive, he raised his voice and heaved at the mirror, determined to flay the guy under that trap door.

Among his first toys I gave him a small rectangular bell that had formerly belonged to a cat. He quickly learned to imitate its flat, metallic sound. And he became very possessive toward it, getting quite upset if anyone threatened to take it away. It served as an important prop in some of his most exciting games. Depending upon his mood, he jabbed it the way a boxer uses a punching bag, all the while screeching insults; or he touched it with the tip of his beak, dreamily warbling.

His most spectacular performance began with a furious whack at the bell and raucous abuse, followed by lively bobbing of his head. Then under the bell he ducked, side stepping to the opposite

Pretty Plaything

end of the perch where he repeated his bowing and chatter. Sometimes the dumb old bell irritated him to a frenzy and he reached up to bite the small clapper.

From afar in the house I could hear the ding ding of Dickey's little bell peremptorily ringing. At first I imagined a note of alarm. It was just possible that some eager cat's eyes were fixed on him with a stare altogether unsettling. But Dickey's bravery was monumental. When one of those monstrous, ferocious creatures circled his cage, he kept his eyes on it, but he didn't panic. And if the cat simply sat close by, he went on with his own affairs, calmly indifferent to darting feline glances. When Mai tired of his antics, she fell asleep leaning against his cage.

Dickey was a chatty soul. One of his favorite pastimes was his own version of talking on the telephone. Facing into a corner of his cage, he held long conversations with an invisible friend. Some of his utterances were very melodic, but mostly he sounded like Victor Borge's famous phonetic pronunciation.

Like any fond mama, I wanted our boy to express himself well. So when he got somewhat settled, we started his summer school. Both of my books on parakeet training stressed patience and persistence, and blamed teacher if the bird failed to talk.

At first we concentrated on, "Hello, Charlie," anticipating the boss's pleased surprise at hearing himself addressed by name. But the time came, seeing that we were totally unsuccessful with this, when I was willing to settle for just plain "Hello." I said, "Hello" to him first thing in the morning, and "Hello" everytime I approached his cage, and "Hello" when we retired to the bathroom for his daily flying exercises. He never returned the courtesy. I changed his lessons from evening to morning hours, thinking that my pupil was perhaps too tired to learn after an exhausting day. No better. Something must be lacking with teacher.

He very politely stepped onto my finger when I offered my hand in his cage. With a little shuffling of his feet and restrained eagerness he waited until I almost touched him, then over he stepped and out he rode. Dickey was a good listener, giving me all his bright-eyed attention, but as a learner he was a bust. He flunked summer school.

I thought then, if he ever does talk, I'll be struck dumb. And if he turns out to be mama's idiot boy, for like one's own "blood baby" we didn't pick him out, but took what we got, then we will console ourselves with the conviction that random chatter has its drawbacks too.

Grace Personified

Meanwhile, we loved him just as he was. Dour wintry mornings the sunshine of his presence and his song lifted the spirit. Very early we found ourselves quite prepared to admire anything the little bird might do. Even when he was very rude. Or naughty. Thinking to entertain him, I whistled a Strauss waltz. He yawned in my face! And he was so cute when he yawned. If he was not in the mood for kisses, he would nip me with his strong little beak. My stern scolding only elicited a bright impish look too comical for gravity.

He and I were not long in establishing such rapport as permitted regular little love feasts before his daily exercise. He knew perfectly what was up when we climbed the stairs. In the bathroom, the door securely latched against feline intrusion, I would invite him out of the cage. Onto my hand he hopped, and up my arm he sidled until he perched on my shoulder. I made kissing noises. His head feathers crested, he repeated the sounds. Over he leaned and gave me nibbly little kisses on the lips. Or he pressed his head against my cheek, wooing me with exquisite trills, delicate and dreamy. Only superhuman indifference could withstand such blandishments!

A bird flying free throughout the house might strain to the breaking point our cats' mistress-imposed forbearance. But in our large bathroom Dickey could be left alone as he pleased for his

daily exercise. Back and forth he flew with great whirring of small wings and much contented twittering. Those periods varied according to his whim, but when he was finished, he returned to his cage and rang his little bell.

Occasionally Dickey tickled us with a tentative wolf whistle. And during his first summer storm, after an ear-splitting crash of thunder, he went, “Whew!”

Through the wintertime Dickey spent his days on the top of our teakwood chest in the dining room. Mai and Araminta liked to sit there too. They watched him, the starry blue eyes and the glittering yellow, glances darting as he moved, while he unconcernedly busied himself about his own affairs. Rubbing bills with his roly-poly penguin, he chirped and spit and twittered; he rocked it and rolled it and sashayed about with fancy footwork. He would rush up for a peck at his bell, then down again for a word with his handsome double in the mirror.

Mai crouched nearby, gazing soulfully at him. If he approached her side of the cage, her whiskers twitched. Before long, however, the sameness of his performance bored her. His tireless activity wore her down, she grew drowsy. Settling herself against the cage, she napped. Dickey’s energy was unflagging. He chirped and warbled and rang his bell in carnival spirit. Mai, curled into a ball, slept peacefully, undisturbed by all the racket.

Most of the day our lad was full of song and fury. He chirped and trilled joyfully; he screeched and scolded. Sometimes his outbursts sounded like hysterical laughter at what must have been a rare story. Off color no doubt. There was quiet only when he was busy hulling his seeds, which he did with marvelous sleight of beak, or when he plumed his feathers. Such lulls were infrequent however. It was “agin’ nature” for Dickey to be mute during waking hours.

At first, before I knew better, we kept our small friend up too late at night. And so during the warm afternoons he would take a nap. Sitting in his swing he fell asleep leaning against one of its supports, his head nodding lower and lower. Such a drooping small bundle of feathers, and so unnaturally quiet, he made a pathetic figure.

Or he stood on one foot, the other doubled into a fist like some poor bony, bloodless hand and skirted in his soft yellow feathers, his beak tucked back under one wing and eyes closed, sleepily warbling to himself. If anyone happened by, he cocked an eye above his buttercup plumage, then resumed his private concert.

Devil of a Fix

Learning that birds should go to bed not long after sundown, I provided the seclusion he needed at the proper hour.

Our lad kept very well, which gave great satisfaction to a greenhorn parent. Mostly he ate what he should, except for lettuce or clover or dandelion, which he detested. But he seemed not to need such dietary supplements. And it is best to treat each pet as an individual. He enjoyed his seed tree. And he nibbled his egg wafer ravenously, holding it to him with one foot while he chipped away until he had a hole in the yellow square that framed his yellow head.

Dickey might not talk (so who cares, everybody talks), but he was something of a musical prodigy. He performed with considerable virtuosity on his bell. His face buried, he pecked the clapper — ting, ting . . . ting, ting — repeated with nuances of expression, interspersing his own

trills and chirping, and climaxing with a crashing crescendo. He did quite a concert.

To further his musical education we bought for him a tuneful perch. This consisted of a music box in a small peaked-roof house of yellow plastic with a round mirror in the front, and below, a short dowel extending from a wire that tripped the mechanism when the bird alighted. A selection of melodies had been available, including a *Viennese Waltz, Brahms' Lullaby* and *Rock-a-bye-Baby.* It was generally agreed that measured waltzes and quiet lullabies were too slow for our exuberant little bird. Proper for him was something like *For He's A Jolly Good Fellow.*

The new instrument intrigued Dickey. He eyed it and listened to the merry air. He examined the small yellow house and admired himself in the mirror. He pecked at the thin dowel and wiggled it with his beak until within the hour he had started the music all by himself. That being fun, he tripped it a second time.

A few days later he played it six times in ten minutes, but not in the prescribed fashion by stepping onto the hop-a-tune. Rather, he pumped it with his beak. We were puffed with pride, watching him figure it out, touching it, shaking it, and fetching the tune. Sometimes when he was in the mood for music and found his box run down, he walloped the trigger, pumping furiously — wham, wham, wham! And a ready hand quickly wound it up for him.

From solo portrait sittings, Dickey graduated to posing with our bird visitors. These began to arrive shortly after he celebrated his first anniversary with us. And they continued to come in a steady stream, eighteen parakeets in a twelve-month period, all colors, all ages from two years to past eleven. I was constantly furnishing new, clean, interesting quarters for birds that were discarded, unappreciated, neglected.

Dickey loved company. He practically turned himself inside out trying to make friends with an apathetic house guest who sat morosely on his perch, responding only with a frog's croak. So exuberantly sociable was Dickey that he overwhelmed these formerly cloistered creatures who had spent their bleak lives unloved. He welcomed them all as kissin' cousins. When he sidled along the perch toward them with excited bird talk, they backed away. But he pressed on with his endearments, sometimes reaching out one foot and clutching the reluctant one to him.

For a decade, Dickey's gay spirit cheered us the day long. Then one afternoon his capers seemed to weary him. He fell asleep huddled in his swing. When his silence lasted overlong. I peeked in. Our little bird was dead. At that moment earth's sunshine went under a cloud.

Barred Owl

Afterthoughts

Devotion to animals can become a bitter-sweet addiction. Sweet in the love we cherish for the wonderful creatures who share our world; bitter for the anguish we feel witnessing cruelties great and small. Our compassion is something of a curse really. For one sensitive to animal woe, it seems almost impossible to venture anywhere without distress over some troubling scene that disturbs indifferent people not at all.

A happy trip to the country is quickly spoiled by the passing glimpse of a poor lonesome dog chained to a doghouse and perched on its roof, eagerly hoping for friendly notice. Animals ask for so little.

Even after thirty years I am still haunted by the scrawny kitten desperately trying to climb the gangplank of an American cruise ship docked at Alexandria, Egypt.

Fortunately for my peace of mind, I could help the injured pigeon huddled on a downtown sidewalk. No telling how long the crumpled bird had lain there; people simply walked around or stepped over it. And the clerk in a nearby store where I asked for a box or paper bag was annoyed: "Why bother? It's just an old pigeon." I bothered because I wanted to sleep that night. The suffering creature needed to be taken where it could be doctored or painlessly put out of its misery. By chance I had been elected to serve.

Happier was the rescue of a handsome barred owl observed sitting day after day on the same limb of a towering oak. Finally two boys climbed up and brought him down. At the vet's it was found that he had a broken wing. While convalescing he spent his days in a bird lover's rec room, fascinated by the antics of little sparrows at a bird feeder outside his window. And even light drinkers, turning away from the bar and noticing him in a shadowy corner, thought that they must be seeing things.

To avoid a warped personality, one should treasure the happy experiences.

Indifferent, skinflint owners got rid of a gentle big white bunny because he was sick. A lady who just happened to stop by the shelter bounced him on her knee, said, "Oh, him has a bad cold," and took the lovable creature home for doctoring.

Black-eyed Susan

Sweet revenge was mine when my photos in court brought a conviction of extreme cruelty for the owner of an abused Cocker Spaniel. The dog's coat had gotten so matted and filthy that he could hardly move, and his neck was circled by an open sore an inch wide where a tethering rope had sawed into the flesh. When he recovered, though human beings hardly deserved it, the dog loved everybody. And he was eager to give me big wet kisses. As tactfully as possible I explained to his new owner that I was not partial to being slobbered on by man or beast.

And there was my timely rescue of a miserable, bedraggled kitten hideously abused by two little hoodlums. She seemed half dead, and recovery was slow, but happily the experience left no lasting scars. Her mother was a Virginia sex kitten, her father a traveling Siamese. We named the little part Oriental Mai Ling.

During my stint as recruiting officer for the U. S. Civil Service Commission in WW II, I once referred as Clerk-Typist a homely Chinese girl, admittedly not too well qualified, yet who might serve to ease the War Department desperate need for clerical help. Came the reply of a wag in the

Personnel Office: "Can't use fish and rise type. Please send Lotus blossom type."

Mai Ling was the Lotus blossom type — lovely throughout her long life. Cruelty can destroy so much beauty.

And a true story, more than twice told, for we liked to brag, boasted about the incredible bravery of a dog named Hugo. He was dragged into the Shelter office by a smart aleck who snapped, "Here, you take him. He's a stupid damn mutt."

We looked down into the sad eyes of an unkempt German Shepherd and could hardly sign him in soon enough.

Good food and proper grooming somewhat restored Hugo's interest in living, but his response to people was merely polite. Then one red-letter day he and a pretty teenager fell in love. In the bosom of his new family, Hugo was admired and pampered. And he was miserable if his adored one happened to be out of his sight. Naturally he was on hand when she attempted to ride a fractious horse. Suddenly the animal reared and threw her. As she was about to be trampled, her family could only watch, frozen in horror. But a "stupid damn mutt" named Hugo rushed in, distracted the animal, and saved his loved one's life.

I have preserved my treasured recollections to intrigue other animal lovers with the beauty and nobility of the wonderful creatures who share our world. And perhaps I can impress upon the animal haters, or those merely indifferent, how much joy they are missing. Like the gentleman in Texas who, finding "Saturday Night Bath" in his PSA Journal, wrote that the picture amused him through a serious operation. And remember the camera club member who said, "I don't care for animals, but I like yours"?

Perhaps I have fulfilled my destiny.

Tickle Toes

Personal Notes

Animals have figured largely in my life since my third birthday when I was given a little black dustmop with shoe button eyes, called Peeleg.

A few years later, as we were about to move away from town, I was persuaded to trade my beloved Collie for a snow-white pigeon. Of course, the pigeon flew back home, leaving me with nothing. That betrayal I have never forgiven.

Between my seventh and twelfth years, I lived on a farm in Maryland. My personal pets included Mollie, a tiger cat abandoned on our shore, a sociable biddy named Mable, and a roly-poly black pony. Dixie I drove to school hitched to a parasoled buggy.

At present I mourn the passing of beautiful, loving Delilah. She was a Tortie and White with green eyes. Wherever we rested together, she always had to be touching me. Her death from cancer shortly after my husband passed away left me truly bereft.

In the course of a telephone conversation about something totally foreign, a friend mentioned that a waif cat was then living in a precarious situation with a couple of toughs. Since events in my life have persuaded me to consider seriously the idea of predestination, I accepted her remarks as an omen that this little creature was meant for me. She was described as orange with amber eyes, a fearful stray who had been frightened as she tried to emerge from wintry woods and a snowball was heaved at her.

Though she lived miles away, I went to adopt her. She was not orange, nor were her eyes amber. Had I seen her in a line-up, I would never have given her a second glance. Her coat is pied — hints of wine splashed through gray tabby, her feet are too big for a girl and her back legs are knock-kneed. Worst of all, she is not very affectionate. But Fate decreed that she is mine. And since I am superior in wisdom and experience, it is up to me that we make a success of our marriage.

The Last Laugh